BRITISH RAILWAYS
STEAMING
THE EX.LNER LINES

Volume Four

Compiled by
PETER HANDS

DEFIANT PUBLICATIONS
190 Yoxall Road,
Shirley, Solihull,
West Midlands

Printed on behalf of Richard Netherwood Ltd., by Gorenjski tisk p.o., Kranj, Slovenia

CURRENT STEAM PHOTOGRAPH ALBUMS AVAILABLE
FROM DEFIANT PUBLICATIONS

BRITISH RAILWAYS STEAMING THROUGH THE SIXTIES

VOLUME 14
A4 size - Hardback. 96 pages
-178 b/w photographs.
£14.95 + £1.50 postage.
ISBN 0 946857 40 7.

BRITISH RAILWAYS STEAMING THROUGH THE SIXTIES

VOLUME 15
A4 size - Hardback. 96 pages
-178 b/w photographs.
£16.95 + £1.50 postage.
ISBN 0 946857 52 0.

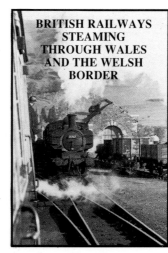

BRITISH RAILWAYS STEAMING THROUGH WALES AND THE WELSH BORDER

A4 size - Hardback. 96 pages
-175 b/w photographs.
£17.95 + £1.50 postage.
ISBN 0 946857 56 3.

BRITISH RAILWAYS STEAM HAULED PASSENGER TRAINS IN THE SIXTIES

VOLUME 1
A4 size - Hardback. 96 pages
-177 b/w photographs.
£14.95 + £1.50 postage.
ISBN 0 946857 41 5.

BRITISH RAILWAYS STEAMING THROUGH THE FIFTIES

VOLUME 9
A4 size - Hardback. 96 pages
-177 b/w photographs.
£14.95 + £1.50 postage.
ISBN 0 946857 37 7.

BRITISH RAILWAYS STEAMING THROUGH THE FIFTIES

VOLUME 10
A4 size - Hardback. 96 pages
-176 b/w photographs.
£14.95 + £1.50 postage.
ISBN 0 946857 38 5.

BRITISH RAILWAYS STEAMING THROUGH THE FIFTIES

VOLUME 11
A4 size - Hardback. 96 pages
-176 b/w photographs.
£16.95 + £1.50 postage.
ISBN 0 946857 48 2.

BRITISH RAILWAYS STEAMING THROUGH THE FIFTIES

VOLUME 12
A4 size - Hardback. 96 pages
-176 b/w photographs.
£16.95 + £1.50 postage.
ISBN 0 946857 49 0.

BRITISH RAILWAYS STEAM HAULED PASSENGER TRAINS IN THE FIFTIES

VOLUME 1
A4 size - Hardback. 96 pages
-177 b/w photographs.
£14.95 + £1.50 postage.
ISBN 0 946857 39 3.

BRITISH RAILWAYS STEAM HAULED FREIGHT TRAINS 1948–1968

VOLUME 1
A4 size - Hardback. 96 pages
-174 b/w photographs.
£14.95 + £1.50 postage.
ISBN 0 946857 42 3.

BRITISH RAILWAYS STEAMING THROUGH THE MIDLANDS

VOLUME 1
A4 size - Hardback. 96 pages
-179 b/w photographs.
£15.95 + £1.50 postage.
ISBN 0 946857 43 1.

BRITISH RAILWAYS STEAMING ON THE EX-LNER LINES

VOLUME 3
A4 size - Hardback. 96 pages
-183 b/w photographs.
£15.95 + £1.50 postage.
ISBN 0 946857 44 X.

FUTURE STEAM PHOTOGRAPH ALBUMS
AND OTHER TITLES

BRITISH RAILWAYS STEAMING ON THE WESTERN REGION

VOLUME 4
A4 size - Hardback. 96 pages -177 b/w photographs.
£15.95 + £1.50 postage.
ISBN 0 946857 46 6.

EARLY AND PIONEER DIESEL & ELECTRIC LOCOMOTIVES OF BRITISH RAILWAYS

A4 size - Hardback. 96 pages -177 b/w photographs.
£15.95 + £1.50 postage.
ISBN 0 946857 45 8.

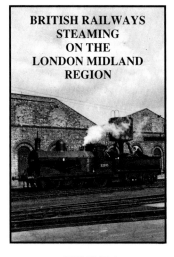

BRITISH RAILWAYS STEAMING ON THE LONDON MIDLAND REGION

VOLUME 4
A4 size - Hardback. 96 pages -177 b/w photographs.
£15.95 + £1.50 postage.
ISBN 0 946857 47 4.

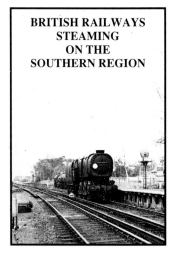

BRITISH RAILWAYS STEAMING ON THE SOUTHERN REGION

VOLUME 3
A4 size - Hardback. 96 pages -177 b/w photographs.
£17.95 + £1.50 postage.
ISBN 0 946857 54 7.

BRITISH RAILWAYS STEAM HAULED TITLED TRAINS

A4 size - Hardback. 96 pages -169 b/w photographs.
£16.95 + £1.50 postage.
ISBN 0 946857 51 2.

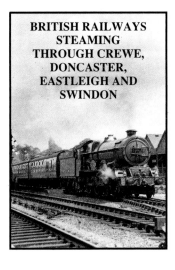

BRITISH RAILWAYS STEAMING THROUGH CREWE, DONCASTER, EASTLEIGH AND SWINDON

A4 size - Hardback. 96 pages -179 b/w photographs.
£17.95 + £1.50 postage.
ISBN 0 946857 53 9.

BRITISH RAILWAYS STEAMING THROUGH LONDON

A4 size - Hardback. 96 pages -174 b/w photographs.
£17.95 + £1.50 postage.
ISBN 0 946857 55 5.

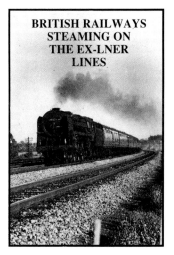

BRITISH RAILWAYS STEAMING ON THE EX-LNER LINES

VOLUME 4
A4 size - Hardback. 96 pages -183 b/w photographs.
£17.95 + £1.50 postage.
ISBN 0 946857 57 1.

BRITISH RAILWAYS STEAMING FROM 1948–1968

'50th' ALBUM
A4 size - Hardback. 96 pages -186 b/w photographs.
£16.95 + £1.50 postage.
ISBN 0 946857 50 4.

BRITISH RAILWAYS STEAMING ON THE LONDON MIDLAND REGION

IN PREPARATION

VOLUME 5

BRITISH RAILWAYS STEAMING ON THE WESTERN REGION

IN PREPARATION

VOLUME 5

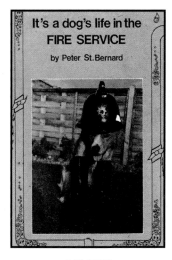

It's a dog's life in the FIRE SERVICE
by Peter St.Bernard

COMEDY
269 pages. Cartoons.
£9.95 + £1.00 postage.
ISBN 0 946857 30 X.

ACKNOWLEDGEMENTS

Grateful thanks are extended to the following contributors of photographs not only for their use in this book but for their kind patience and long term loan of negatives/photographs whilst this book was being compiled.

G.D.APPLEYARD MIDDLESBROUGH	T.R.AMOS TAMWORTH	ALAN BAILEY LEEDS
W.BOYDEN BEXHILL	B.W.L.BROOKSBANK MORDEN	N.L.BROWNE ALDERSHOT
R.BUTTERFIELD MIRFIELD	R.S.CARPENTER BIRMINGHAM	KEN ELLIS SWINDON
A.N.H.GLOVER BIRMINGHAM	S.GRADIDGE CHALFONT ST.GILES	B.K.B.GREEN WARRINGTON
D.HARRISON CHAPELTOWN	PETER HAY HOVE	R.W.HINTON GLOUCESTER
I.J.HODGSON CAMBRIDGE	H.L.HOLLAND ST.CATHERINES ONTARIO, CANADA	F.HORNBY NORTH CHEAM
A.C.INGRAM WISBECH	D.K.JONES MOUNTAIN ASH	M.JOYCE HITCHIN
R.J.LEITCH SAWSTON	ERIC LIGHT TICKHILL	A.F.NISBET BRACKLEY
D.OAKES HITCHIN	R.PICTON WOLVERHAMPTON	A.J.PIKE *
N.E.PREEDY GLOUCESTER	P.A.ROWLINGS ALCONBURY	J.SCHATZ LITTLETHORPE
K.L.SEAL ANDOVERSFORD	J.M.TOLSON BIGGLESWADE	D.WEBSTER **

* Courtesy of the Frank Hornby collection.
** Courtesy of the Norman Preedy collection.

Front Cover - Work-stained BR *Britannia* Class 4-6-2 No 70039 *Sir Christopher Wren*, from 40B Immingham, hammers past Potters Bar with a Kings Cross to Cleethorpes express in June 1962. Once of 30A Stratford and 32A Norwich, No 70039 was drafted to the London Midland Region at 12B Carlisle (Upperby) in December 1963. It survived in revenue earning service in the Carlisle area until September 1967. (N.E.Preedy)

ISBN 0 946857 57 1

© PETER HANDS 1996
FIRST PUBLISHED 1996

INTRODUCTION

BRITISH RAILWAYS STEAMING ON THE EX.LNER LINES - Volume Four is the fourth book to concentrate on the now British Railways tracks and locomotives once owned or influenced by this once great railway company. The author hopes the reader will enjoy the diverse variety of locomotives and locations within the pages of this album.

The 'BR Steaming' books are designed to give the ordinary, everyday steam photographic enthusiast of the 1950's and 1960's a chance to participate in and give pleasure to others whilst recapturing the twilight days of steam.

Apart from the 1950's and 1960's series, individual regional albums like this one will be produced from time to time. Wherever possible, no famous names will be found nor will photographs which have been published before be used. Nevertheless, the content and quality of the majority of photographs used will be second to none.

BRITISH RAILWAYS STEAMING ON THE EX.LNER LINES - Volume Four is divided into three chapters covering the Eastern, North Eastern and Scottish Regions of British Railways from 1948-1967, by which time allocated steam had finished on all three regions. Unless otherwise stated all locomotives are of LNER origin.

The purists might argue that not all of the locations and locomotives included in this album are of pure LNER origins. The author has included some photographs of areas taken over by the BR regions and of locomotives constructed after nationalisation in 1948, but allocated to the same. The author has attempted to vary the locations as much as possible, but some areas of greater interest e.g. Doncaster, York and Edinburgh etc., have been given more coverage than others.

The majority of the photographs used in this album have been contributed by readers of Peter Hands series of booklets entitled "What Happened to Steam" & "BR Steam Shed Allocations" and from readers of the earlier "BR Steaming Through The Sixties" albums. In normal circumstances these may have been hidden from the public eye forever.

The continuation of the "BR Steaming" series etc., depends upon you the reader. If you wish to join my mailing list for future albums and/or feel you have suitable material of BR steam locomotives between 1948-1968 and wish to contribute them towards this series and other albums, please contact:

Tel. No.
0121 745-8421

Peter Hands,
190 Yoxall Road,
Shirley, Solihull,
West Midlands B90 3RN

CONTENTS
EASTERN REGION

NORTH EASTERN REGION

SCOTTISH REGION

NAMEPLATES - Some example nameplates of L.N.E.R. locomotives

1) A3 Class 4-6-2 No 60052 *Prince Palatine*. (R.W.Hinton)

2) B1 Class 4-6-0 No 61022 *Sassaby*. (S.Gradidge)

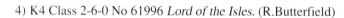

3) B17/1 Class 4-6-0 No 61634 *Hinchingbrooke*.
   (N.L.Browne)

4) K4 Class 2-6-0 No 61996 *Lord of the Isles*. (R.Butterfield)

5) D30 'Scott' Class 4-4-0 No 62438 *Peter Poundtext*.
   (N.L.Browne)

6) The last survivor of the B12 Class 4-6-0's No 61572 takes refreshment as it stands over an ashpit near to a Brush Type 2 diesel at its home base of 32A Norwich on 14th May 1961. Once of 32B Ipswich, No 61572 had been based at Norwich shed since October 1959. After withdrawal in September 1961 it was stored at 30A Stratford and various other locations before being preserved by the North Norfolk Railway. (N.E.Preedy)

7) With a large building dominating the skyline, Thompson inspired L1 Class 2-6-4T No 67788, based locally at 38A Colwick, gently lifts its safety valves whilst running bunker-first with a local passenger train at Nottingham (Victoria) station sometime between late 1957 and early 1958 when the shedcode of Colwick changed to 40E. No 67788, for many years a resident of Colwick, was withdrawn from service in May 1962 and scrapped at Darlington. (R.S.Carpenter)

8) A murky day at 34E New England shed on 1st September 1963 where the massive concrete tower in the background overlooks the depot. Named after the great man himself, A4 Class 4-6-2 No 60007 *Sir Nigel Gresley*, native to 34E, is seen in the company of two LMS Class 4 'Flying Pig' 2-6-0's and a BR Class 9F 2-10-0. During the following month *Sir Nigel Gresley* was drafted to Scotland where it remained until withdrawn for preservation in February 1966. (K.L.Seal)

9) All of the LMS 'London, Tilbury & Southend' Class 3P 4-4-2 Tanks, first introduced into service in 1923, served from sheds on the Eastern Region of British Railways. One example, No 41948, from 33A Plaistow in London and in a quite presentable condition, is noted between duties at Southend-on-Sea station on 11th September 1954. It remained at Plaistow shed until condemned in February 1959 and was later cut up at Cashmores, Great Bridge. (B.K.B.Green)

10) Al Class 4-6-2 No 60157 *Great Eastern* awaits its next working near to the dilapidated wooden coal stage at its home at 36A Doncaster on 8th September 1963. Constructed at Doncaster Works under BR ownership, *Great Eastern* first appeared in 1949 and was one of five examples fitted with roller bearings. Despite having youth on its side it was taken out of revenue earning service in January 1965 and scrapped at Drapers, Hull three months later. (K.L.Seal)

11) With its compact tender packed to the brim with fresh coal supplies and sporting a local passenger headcode D16 Class 'Claud Hamilton' 4-4-0 No 62529 lets off steam from its cylinder cocks as it prepares to leave the yard of its home shed at 31B March on 27th May 1956. By the mid-fifties the writing was on the wall for this famous class and they were rendered extinct by November 1960. No 62529 went in December 1959. (R.Butterfield)

12) This is an unusual profile of A3 Class 4-6-2 No 60044 *Melton*, from 34A Kings Cross, seen at Hitchin on a soaking wet and dismal day in 1962. Built at Doncaster in 1929 its original number as applied by the LNER was 2543. Later modifications included a double chimney in October 1959 and German style smoke deflectors in August 1961. Withdrawn in June 1963 from Kings Cross shed, *Melton* was cut up at its birthplace at the end of 1963. (D.Oakes)

13) The Ll Class 2-6-4 Tanks were subjected to the mass withdrawals of 1962 and on 18th November 1962 No 67745 is only a matter of weeks away from condemnation as it stands out of steam in the yard of its home shed at 40E Colwick. Once of 34D Hitchin and 34A Kings Cross, No 67745 was transferred to Colwick in November 1961. After withdrawal it lay rotting and unwanted until called to Darlington Works for scrapping in March 1963. (J.Schatz)

14) Two elderly locomotives grace the shed yard scene at 31D South Lynn, a shed of Midland & Great Northern Railway origins. Nearest the camera is J66 Class 0-6-0T No 68378 from a type first introduced in 1886. Standing alongside No 68378 is J5 Class 0-6-0 No 65481 on 4th May 1952. Both locomotives were destined for withdrawal and oblivion well before the fifties came to a close. South Lynn 'officially' closed in February 1959. (A.N.H.Glover)

5) The small London Transport Authority terminus at Ongar in Essex hosts a rather less than clean F5 Class 2-4-2T No 67193, complete with a stovepipe chimney, on 26th May 1957. This locomotive is in charge of a pair of ancient carriages providing a local service to Epping, via Blake Hall and North Weald. Allocated to 30A Stratford, No 67193 was rendered surplus to requirements by the operating authorities in November 1957. (N.E.Preedy)

6) In desperate need of the cleaners attentions at 41H Staveley GC, 01 Class 2-8-0 No 63670 rattles a lengthy Class 8 loose-coupled freight train towards the camera on quadruped track at Barnetby, Lincolnshire on 21st November 1963 on its way from Immingham to Scunthorpe. Once of 38B Annesley and 31B March, No 63670 had moved to 41H in February 1960. Condemned in June 1964 it was later despatched to Wards, Killamarsh for scrapping. (N.E.Preedy)

17) A feather of steam escapes from the safety valves of Al Class 4-6-2 No 60124 *Kenilworth* as pressure builds up in the boiler as it rests briefly at St.Neots station with an unidentified working on 23rd June 1963. Based at 50A York, *Kenilworth* had previously been allocated to 52A Gateshead and 52B Heaton. Drafted to 51A Darlington in December 1964 it became the penultimate survivor of the class being withdrawn in March 1966. (R.J.Leitch)

18) The 'old' and 'new' order at 32B Ipswich on 5th July 1959. 31B March based B17 Class 'Sandringham' 4-6-0 No 61660 *Hull City*, in reasonable condition, is prepared for the road amidst the diesels. Withdrawn from 31B in June 1960 this was the only member of the class which the author saw, in store at 2A Rugby during the same year. It was cut up at Stratford Works in August 1960, the same month that the last working example was withdrawn. (N.E.Preedy)

9) The J11 Class 0-6-0's, designed by Robinson and first introduced during 1901, were widely employed on freight, local passenger, station pilot and shunting duties until their demise in 1962 and were based at a wide variety of depots on the Eastern Region of British Railways. On an unknown date in the early fifties No 64283 is a visitor to 36A Doncaster from 36E Retford. Condemned in April 1959 it was scrapped at Gorton Works. (D.K.Jones)

10) Under clear signals BR Class 9F 2-10-0 No 92144, from 34E New England, passes the small north signalbox and trundles a lengthy loose-coupled Class 8 freight through Peterborough (North) station on 10th August 1960. Entering service at New England shed in August 1957, No 92144 remained there until September 1962 when it was moved to 40B Immingham. A few months later it was back at 34E. It later served at 4IJ Langwith Junction and 40E Colwick. (D.K.Jones)

21) The Staveley Iron Works complex was the saviour for a number of locomotives from the LMS 0F 0-4-0 Tanks, 1F 0-6-0 Tanks and 1F 0-4-0 Saddle Tanks helping to prolong their active lives well beyond their 'sell-by date'. In this photograph, No 41763 is seen at work on shunting duties at the complex in October 1958. This engine was not withdrawn until December 1966 and after a period of storage it was cut up at Arnott Young of Parkgate. (N.E.Preedy)

22) Former Great Central Railway freight engine power on show in part of the large shed yard at 36C Frodingham (Scunthorpe) on 24th July 1955. In the left of the frame is 04/8 Class 2-8-0 No 63793 (withdrawn in May 1965). To the right of No 63793 is sister engine 04/7 No 63747 (withdrawn in May 1961). Both engines are local to Frodingham. The shed housed a host of these locomotives prior to closure to steam in February 1966. (R.Butterfield)

3) An immaculate looking 1947 built A2/3 Class 4-6-2 No 60523 *Sun Castle* is a visitor to 34E New England from 34F Grantham in March 1959 one month before being transferred to 34E. In January 1960 *Sun Castle* was transferred to 36A Doncaster, but returned for a final stint at New England prior to being taken out of traffic in June 1963. Behind *Sun Castle* is K3 Class 2-6-0 No 61978, a local inhabitant of New England shed. (A.C.Ingram)

24) A busy scene within the gloomy confines of Nottingham (Victoria) station on an unknown date in the fifties. Note the dilapidated roof structures in the background. In the foreground the fireman of K2 Class 2-6-0 No 61777 (38A Colwick) sorts his coal supplies out prior to setting off with a local passenger train. Standing adjacent to No 61777 is B1 Class 4-6-0 No 61188, also of Colwick, with another local passenger working. (R.W.Hinton)

25) It is raining 'Cats and Dogs' at Peterborough (North) station on 30th March 1963, where we espy a Gresley Pacific nearing the end of its illustrious career. 34F Grantham based A3 Class 4-6-2 No 60106 *Flying Fox*, fitted with a double chimney in April 1959 and German style smoke deflectors in October 1961, is in charge of a down semi-fast passenger from Kings Cross. *Flying Fox* survived in service until December 1964. (D.K.Jones)

26) The LMS Hughes Class 6P5F 'Crab' 2-6-0's were no strangers to Eastern Region metals, but quite what No 42897, from 16D Annesley, is doing at Littlefield, near Grimsby in early June 1960 is not known as it trails along light engine. Later this same month No 42897 was reallocated to 17B Burton. In November of the following year it was drafted to 16A Nottingham where it remained until no longer of use in April 1964. (D.K.Jones)

27) A mixture of steam engine classes await their next duties in the shed yard at 36B Mexborough in February 1957. Nearest the camera is J11 Class 0-6-0 No 64285 which was withdrawn from Mexborough in August 1958. It was cut up by BR at Gorton Works in February 1959. Next to No 64285 is 04/8 Class 2-8-0 No 63828, another Mexborough engine, which was withdrawn from 4IJ Langwith Junction in August 1965. In the distance is a WD Class 8F 2-8-0. (N.E.Preedy)

28) The LMS 'London, Tilbury & Southend Railway' Class 3F 0-6-0 Tanks which first came into service at the beginning of the twentieth century eventually came to a total of fourteen units in number and their ranks remained intact until April 1958. From then on they were decimated and all were gone by February 1959 with the exception of No 41981 which soldiered on until June 1962. On 2nd August 1955, No 41992 (33A Plaistow), is noted at 33B Tilbury. (B.K.B.Green)

29) A3 Class 4-6-2 No 60054 *Prince of Wales*, from 34E New England, pilots an unidentified B1 Class 4-6-0 onto the down slow line at Sandy as they canter along to Peterborough as light engines in April 1964. Withdrawn two months later, *Prince of Wales* had been in service for some forty years. Modified by BR with a double chimney and German style smoke deflectors like most of its sister engines, No 60054 was cut up at Kings, Norwich. (I.J.Hodson)

30) With a fine array of lower quadrant signals in the left background, E4 Class 2-4-0 No 62785, based locally at 31A, poses for the camera at Cambridge on 16th May 1959. Designed by J.Holden for the Great Eastern Railway, No 62785 was constructed at Stratford Works in 1895 and withdrawn from 31A in December 1959. All in all a total of 100 of these engines were built between 1891 and 1902 and they were the last 2-4-0's in service. (F.Hornby)

31) The diminutive two-road shed at the outpost of Melton Constable, coded 32G under British Railways, originally belonged to the Midland & Great Northern Railway and after nationalisation came under the ownership of the Eastern Region. At any one time it only had a small allocation and in 1951 had to be rebuilt in brick after the original shed collapsed. It closed in February 1959. On view on 19th September 1953 is F6 Class 2-4-2T No 67224. (A.N.H.Glover)

32) Especially spruced-up for the occasion, B1 Class 4-6-0 No 61131, from 56A Wakefield, is noted at Shireoaks on 13th August 1966 whilst in charge of the Railway Correspondence and Travel Society inspired 'Great Central Tour'. From 1957 until condemnation in December 1966, No 61131 was allocated to 56A Wakefield (twice), 56B Ardsley (twice), 56F Low Moor and 56G Bradford (Hammerton Street). It was scrapped at Drapers, Hull in 1967. (N.E.Preedy)

33) A filthy dirty LMS Class 4 'Flying Pig' 2-6-0 No 43109, allocated to 40F Boston, issues black smoke as it struggles to find its feet whilst in the process of hauling a dead unidentified V2 Class 2-6-2 out of Whitemoor marshalling yard, March towards the nearby shed at 31B on an unknown day in September 1960. From the late fifties until withdrawal in November 1965, No 43109 served from a number of sheds on the Eastern Region. (N.E.Preedy)

34) The powerful former Great Northern Railway 02 Class 2-8-0's were untouched by withdrawals until 1960, but within a few short years they were rendered extinct. During the latter part of their lives all sixty-six engines were based at Doncaster, Grantham and Retford sheds. Locally based No 63942, in a somewhat less than pristine condition, is photographed in the yard at 36E Retford GC on 27th August 1961 two years before condemnation. (T.R.Amos)

35) A parade of neglected steam locomotives at rest over the ashpits outside the motive power depot at 34E New England on 1st September 1963, a gloomy and wet day. On the left is an unidentified B1 Class 4-6-0 which is standing next to A4 Class 4-6-2 No 60026 *Miles Beevor* which was destined to be transferred to Scotland the following month. The other two locos are BR Class 9F 2-10-0 No 92141 and A1 Class 4-6-2 No 60121 *Silurian*. (K.L.Seal)

36) BR Class 9F 2-10-0 No 92186, from 40E Colwick, rattles a southbound Class 7 partially fitted freight through Grantham station on a sun-filled 26th August 1964. Entering traffic in January 1958 and equipped with a double chimney, No 92186 was allocated to 35A/34E New England where it remained until June 1963, moving to Colwick. During the final year of its short life it was based at 36A Doncaster and 41J Langwith Junction. (K.L.Seal)

37) Despite still carrying its coal supplies it is the end of the road for former Great Central Railway 01 Class 2-8-0 No 63678 as it lies unwanted in a neglected condition in the shed yard at 41H Staveley GC on 27th July 1963 after withdrawal. Once of 38B Annesley and 31B March, this locomotive had been transferred to the ex. GC shed at Staveley in February 1960. Two months after withdrawal it was cut up at Doncaster Works. (N.E.Preedy)

38) 31B March hosts a visitor in the shape of LMS Stanier Class 6P5F 2-6-0 No 42975, from 2B Nuneaton, on 16th September 1962 which has been serviced and ready for the road. Out of a total of forty engines the class was rendered extinct by February 1967, with the exception of No 42968 which is now preserved in active condition on the Severn Valley Railway. No 42975 was withdrawn from service in March 1966 from 9F Heaton Mersey. (T.R.Amos)

39) Whilst on the subject of preservation, the locomotive in this photograph is the only one of its class to be preserved. N7/4 Class 0-6-2T No 69621, based at 30A Stratford, trundles up the gradient on non-electrified track and enters Bethnal Green station with a lengthy commuter train in September 1959 shortly after being transferred to 30A from 32C Lowestoft (Central). Today, it is preserved on the Stour Valley Railway. (N.E.Preedy)

40) Yet another steam locomotive in neglected condition. Begrimed Thompson B1 Class 4-6-0 No 61159 roars through Welwyn Garden City station on a centre road with an express during the summer of 1962. Based at 40B Immingham, No 61159 is probably on a Grimsby-Kings Cross-Grimsby working. This locomotive was allocated to 40B for many a year and remained there until taken out of revenue earning service in September 1963. It was cut up at Cashmores, Great Bridge in February 1964. (R.W.Hinton)

41) Steam leaks from various places on B1 Class 4-6-0 No 61074, from 34E New England, as it trundles over a boarded crossing leading to a goods shed at Huntingdon North station on 27th February 1962. No 61074 is in charge of a down goods train. This locomotive had been transferred from the North Eastern (53A Hull - Dairycoates) to the Eastern Region (35B Grantham) in April 1957. A few months later it moved on to New England. (T.R.Amos)

42) Locomotives recently outshopped from Doncaster Works are lined up in the shed yard at 36A Doncaster in readiness to be steamed and returned to their home depots. Included in this line-up on 13th October 1957 is LMS Class 4 'Flying Pig' 2-6-0 No 43154, of 32G Melton Constable. When this shed closed in February 1959, No 43154 moved on to pastures new at 40E Colwick. It also served from 40A Lincoln before returning for a final fling at Colwick. (B.K.B.Green)

43) Overlooked by various domestic dwellings it is the end of the line for former Great Central Railway 04/3 Class 2-8-0 No 63798 as it stands lifeless, minus coal supplies on a dead road at its home base of 36E Retford GC on 29th April 1962. Transferred to Retford from 41F Mexborough in August 1961, No 63798 was not 'officially' taken out of service until May 1962, after which it was despatched to Gorton Works for scrapping. (N.E.Preedy)

44) Another locomotive nearing the end of its working life is L1 Class 2-6-4T No 67713, from 31A Cambridge, seen here in this photograph in a begrimed external condition shunting wagons at Bishops Stortford in June 1961. Once of 30C Bishops Stortford itself, No 67713 had been at Cambridge shed since April 1959. Having been rendered surplus to requirements in October 1961 it was sent north to Darlington works for cutting up. (A.J.Pike)

45) Conisborough, near Mexborough is the setting for this print although the spelling on the diminutive signalbox is 'Conisboro'. On a gloomy day in 1960 K3 Class 2-6-0 No 61912 passes the signalbox and extensive colliery with an excursion. No 61912 is based at 40B Immingham and moved on to 40A Lincoln in October 1961 and finally to a last base at 34A Kings Cross in April 1962. It was later used as a stationary boiler at 34A and 34E New England. (R.S.Carpenter)

46) The four-road shed at Kings Lynn, once the property of the Great Eastern Railway, was coded 31C under British Railways. It was closed to steam in April 1959 though it continued to service the odd visitor for several months afterwards. It closed completely in 1961 and was demolished during the same year. Photographed in the shed yard on 4th May 1952 is D16 Class 4-4-0 No 62525 which was withdrawn well before the end of the fifties. (A.N.H.Glover)

7) Escaping steam partially obscures the cabside number of Thompson B1 Class 4-6-0 No 61120 as it rests in ex.works condition on the turntable at 41A Sheffield (Darnall) on a cold but dry day on 27th January 1962. No 61120 is a visitor to this former Great Central Railway depot from 36E Retford. Closed to steam in June 1963, Darnall continued to host the odd visitor until later in the year and also stored steam locomotives. (P.A.Rowlings)

8) Standing amidst piles of discarded ash F5 Class 2-4-2T No 67213, from 30A Stratford, looks in a quite presentable condition in the sidings at Ongar station with two coaches which will later form a local passenger to Epping. The Great Eastern Railway owned a total of 235 2-4-2 Tanks and used them on London suburban services and on many branch lines on the GER system. No 67213 was withdrawn well before the fifties came to a close. (N.E.Preedy)

49) With a pair of sister engines in tandem on a passenger train in the distance, LMS Class 4 'Flying Pig' 2-6-0 No 43150 stands in a platform at Melton Constable station with the stock of the 10.35 am from Peterborough to Yarmouth on 6th August 1955. No 43150, a locally based locomotive at 32G, was transferred to 30A Stratford upon the closure of Melton Constable shed in February 1959. It later served from 34E New England. (B.K.B.Green)

50) Sunlight and shadow at Hitchin station on 11th August 1962. In charge of a Kings Cross to Peterborough local passenger is work-stained A3 Class 4-6-2 No 60046 *Diamond Jubilee*, from 34E New England. Once of 36A Doncaster and 34F Grantham, this 1924 built locomotive had been modified with a double chimney in July 1958 and fitted with German style smoke deflectors in December 1961. It was withdrawn from 34F Grantham in June 1963. (D.Oakes)

1) Looking in fine fettle B1 Class 4-6-0 No 61005 *Bongo* simmers in the yard of its home depot at 30F Parkeston on 25th May 1956. By January 1957 *Bongo* was allocated to 35A New England and during the following month it was drafted to 35C Spital Bridge (Peterborough). After a stint at 30A Stratford, from June 1957 until January 1959, No 61005 was once again back at Parkeston shed. It later served from the depots at 31A Cambridge and 31B March. (N.E.Preedy)

2) Former Robinson Great Central Railway O1 Class 2-8-0 No 63592, from 40E Colwick, looks as though it is still in active service as it stands in begrimed condition in part of the shed yard at 41J Langwith Junction on 27th July 1963, although it was withdrawn during this same month. Also during July 1963 it was subjected to a transfer to 41H Staveley GC, though whether this was a 'paper' transfer is debatable. It was cut up at Doncaster later in the year. (N.E.Preedy)

53) The K3 Class 2-6-0's were designed by Sir Nigel Gresley and came under the power specification of 5P6F and were first introduced into traffic in 1924. They were employed on both passenger and freight services on the Eastern, North Eastern and Scottish Regions of British Railways. On 21st August 1960, No 61811, its tender filled with slack, is a visitor to 36C Frodingham from 41H Staveley GC. It survived in service until October 1962. (N.E.Preedy)

54) Gaunt and grimy buildings form a backdrop to this photograph taken at Nottingham (Victoria) station during the mid-fifties. As a begrimed water column on the platform keeps a silent vigil, J6 Class 0-6-0 No 64256, of 38A Colwick, waits patiently with a local passenger working. In October 1957, No 64256 departed from Colwick shed, moving to 38C Leicester GC. It ended its working life based at 9G Gorton, being condemned in May 1960. (R.S.Carpenter)

55) With a virtually empty station in the left background, one of the legions of Riddles War Department Class 8F 2-8-0's, No 90501, based at the local depot (31B), departs from March station with a loose-coupled freight in August 1960. Apart from 31B, No 90501 worked from a host of sheds from January 1957 until withdrawn in November 1965, including 35C Peterborough (Spital Bridge), 30A Stratford, 36A Doncaster and 36C Frodingham. (N.E.Preedy)

56) Allocated to Sheffield (Darnall) shed, B1 Class 4-6-0 No 61152 is in immaculate external condition as it poses for the camera near to the running shed at 38E Woodford Halse on 25th March 1954. From 1949 to 1955, Darnall was coded 39B, after which it was recoded 41A until closure. Woodford Halse became the property of the London Midland Region in February 1958 and was coded 2G, 2F and 1G before complete closure on 14th June 1965. (N.E.Preedy)

57) Quite what 52A Gateshead based Al Class 4-6-2 No 60143 *Sir Walter Scott* is doing in the depot yard at 34D Hitchin on 7th March 1960 is possibly only something the photographer can shed light on - a failure, perhaps! *Sir Walter Scott* is seen in the company of an unidentified English Electric Bo-Bo Type 1 diesel. Whatever the reason for the visit of No 60143 it was destined to remain in service until May 1964, being withdrawn from 50A York. (M.Joyce)

58   Sporting white express headcode discs, common to the Eastern Region, BR *Britannia* Class 4-6-2 No 70041 *Sir John Moore*, of 30A Stratford, speeds towards the camera in the mid-fifties with a Norwich train near to Colchester. *Sir John Moore* remained on the Eastern Region until December 1963, serving also from 32A Norwich and 40B Immingham. During the last few years of its working life it was based at both sheds in Carlisle. (D.K.Jones)

49) Stevenage station on the East Coast Main Line, some twenty-nine miles from Kings Cross is the setting for this photograph. A4 Class 4-6-2 No 60026 *Miles Beevor*, from 34A Kings Cross, bears down on the photographer as it speedily heads a down express freight from Ferme Park on 25th July 1960. Based at 34A for many a year, *Miles Beevor* moved on to 34E New England in June 1963, followed by a later move north of the border to Scotland. (M.Joyce)

50) Tank engine power on show in the yard at 31A Cambridge on 2nd October 1955. Bearing the mark of 'doom', sacked chimneys, are two stored F6 Class 2-4-2 Tanks, Nos 67227 and 67238. Of these, No 67227 was later returned to traffic, being withdrawn from 30E Colchester in May 1958. Keeping the two 2-4-2 Tanks company is J69 Class 0-6-0T No 68555, also withdrawn in May 1958, latterly of the sheds at 32A Norwich and 32B Ipswich. (R.Butterfield)

61) Spalding and its immediate area was once the junction of lines radiating to many points of the compass and was also a stabling point for steam locomotives when not in use. Standing dead in the sidings at Spalding on 27th May 1951 are two LMS Class 4 'Flying Pig' 2-6-0's, Nos 43087 and 43066. Note the single line tablet catchers by the cabs. No 43087, from 35A New England, survived in service until December 1964 and was cut up in 1965. (B.K.B.Green)

62) Sporting a scorched smokebox door A3 Class 4-6-2 No 60067 *Ladas*, from 34F Grantham, lifts its safety valves as it passes beneath Spital Bridge at Peterborough with a Kings Cross bound express in July 1959. A double chimney had been fitted to *Ladas* three months earlier, later followed by German smoke deflectors in April 1961. This latter modification did little to prolong the life of *Ladas*, being condemned from 34A Kings Cross in December 1962. (A.C.Ingram)

3) Looking in fine external fettle, Peppercorn A1 Class 4-6-2 No 60122 *Curlew*, allocated to 36A Doncaster, pauses at Baldock station with an express on an unknown date in 1960. Built at Doncaster under the auspices of British Railways in 1948, *Curlew* was the only example of this class which the author did not see in his 'spotting' days. It was one of the first members of the class to be withdrawn, from 36A Doncaster in November 1962. (D.Oakes)

4) We take our leave of the Eastern Region section of this album with this shot of a Holden designed J17 Class 0-6-0 No 65521 in light steam in the yard of its home shed at 30A Stratford on a gloomy and misty 8th June 1961. Once of 31C Kings Lynn and 31B March, No 65521 had found its way to 30A Stratford in June 1961. In September of this same year it was drafted to 32A Norwich from whence it was taken out of traffic in February 1962. (D.K.Jones)

65) With its exhaust being blasted high into the sky, LMS Class 4 2-6-4T No 42233, of 56F Low Moor, comes around the curve and passes the site of St.Dunstans station, Bradford with the Bradford portion of the *Yorkshire Pullman* on 15th April 1967. St.Dunstans, once owned by the Great Northern Railway, had closed as far back as 1952. No 42233 had been transferred from the London Midland to the North Eastern Region in November 1966. (N.E.Preedy)

56) The impressive concrete built water tower in the shed yard at 50A York looks down upon Al Class 4-6-2 No 60126 *Sir Vincent Raven*, which, although fully coaled, is out of steam on 6th August 1963. *Sir Vincent Raven*, a local steed, had been allocated to 50A since a move from 52B Heaton in August 1961. Constructed at Doncaster Works in 1949, No 60126 was destined to remain in service until January 1965, being cut up by Drapers, Hull. (H.L.Holland)

57) The state of the locomotive in this photograph does little to enhance the public image of British Railways to the potential traveller. Although it is high summer it is a gloomy and damp day at West Hartlepool station as a young enthusiast 'admires' the presence of an elderly G5 Class 0-4-4T No 67324, a local steed from 51C, which is in charge of a local passenger train on 11th August 1956. Some of these engines were fitted with push-and-pull apparatus. (Peter Hay)

68) A solitary mineral wagon occupies a weed-strewn siding at Longwood on the outskirts of Huddersfield on a hazy day on 17th July 1965 as 9H Patricroft based BR Class 5 4-6-0 No 73033 blasts by with a westbound partially fitted freight. Note the crude 9H stencilled on the smokebox door near to what was the code of 5E (Nuneaton) which was the previous abode of No 73033 prior to July 1965. It survived at Patricroft until withdrawal in January 1968. (D.K.Jones)

69) A profile of J94 Class 0-6-0ST No 68027 seen here in fine external fettle in the yard of its home shed at 51A Darlington on 22nd June 1952. No 68027 spent most, if not all, of its working life based at 51A up until withdrawal in December 1960. This class was designed for the Ministry of Supply by the Hunslet Engine Company and between 1943 and 1946 hundreds were built by private contractors. In 1946 the LNER purchased seventy-five of them. (A.N.H.Glover)

0) A northerly outpost of steam was at West Auckland which had its own motive power depot (51F) until complete closure on 1st February 1964. In June 1960 a trio of local inhabitants are gathered near to the running shed. From left to right are: J39 Class 0-6-0 No 64756 (withdrawn in December 1962), BR Class 4 2-6-0 No 76046 (withdrawn in May 1967 from 67A Corkerhill - Glasgow) and LMS Class 2 2-6-0 No 46471 (withdrawn in November 1962). (Ken Ellis)

71) 52A Gateshead possessed a small stud of A4 Class 4-6-2's over the years and they tended to be the ones which had non-corridor tenders. It was also the last depot in England to house these magnificent engines. One example, No 60001 *Sir Ronald Matthews*, pauses adjacent to Newcastle (Central) station on an unknown date in 1963. This locomotive was destined to be the last surviving member of the class at 52A, being withdrawn in October 1964. (D.K.Jones)

72) The former North Eastern Railway depot at Leeds (Neville Hill) had on its books over the years a number of Al and A3 Pacifics within the confines of its four covered roundhouses, two of which were demolished to make way for a diesel depot in 1958. On 28th July 1962 one of its stud of A3 Class 4-6-2's, No 60084 *Trigo* is seen at rest near to a turntable. *Trigo*, equipped with a double chimney and German smoke deflectors, was withdrawn in November 1964. (D.K.Jones)

73) Another former North Eastern Railway shed was at Hull (Botanic Gardens), a large affair amidst a sprawling mass of tracks leading to and from Botanic Gardens Junction and West Parade Junction. Despite major rebuilding in 1957 it was closed to steam on 14th June 1959, carrying the BR shedcode of 53B up until closure. On 31st May 1953 B1 Class 4-6-0 No 61060 is a visitor to Botanic Gardens from another local shed, 53A Hull (Dairycoates). (A.N.H.Glover)

74) Overhead wires and the associated masts etc., clutter the skyline at Penistone West, situated between Barnsley and Manchester on the ex. Great Central/Lancashire & Yorkshire Joint Railway. In charge of a lengthy excursion on a dismal 7th June 1954, which consists of a mixed bag of stock, is LMS Hughes Class 6P5F 'Crab' 2-6-0 No 42903, from 21A Saltley. Withdrawn from Saltley in September 1962, No 42903 was later cut up at Horwich. (B.K.B.Green)

75) A splendid panoramic view as taken from the unkempt and closed station at York (Holgate) on a sun-filled 19th May 1964. Heading north is a down fitted freight which is in the capable hands of 50A York based BR Class 9F 2-10-0 No 92206 (equipped with a double chimney). In September 1963 a batch of these engines were allocated to 50A, these being Nos. 92005/6, 92205/6/11/21/31/39. All were gone from York by November 1966. (H.L.Holland)

76) The 0-6-0 Wheel arrangement was amongst the most common in Britain and of these one of the most popular classes were the Aspinall designed Lancashire & Yorkshire Class 3F's. 484 were constructed between 1889 and 1918 including No 52515, seen here fresh from overhaul at its home shed of 56D Mirfield on 16th February 1960. This locomotive was destined to be the last survivor, being withdrawn from 56E Sowerby Bridge in December 1962. (D.K.Jones)

77) A usurper on the North Eastern Region in the shape of work-stained LMS Class 5 4-6-0 No 45015, from 8A Edge Hill (Liverpool), seen here at a colour light signal near to the water tower/coaling stage at Normanton shed on a gloomy 5th April 1962. Once of 3B Bushbury (Wolverhampton) for many years, No 45015 was reallocated to 8A in November 1961. It stayed at Edge Hill until rendered surplus to operating requirements in September 1967. (D.K.Jones)

'8) The tranquility of the north-eastern countryside at Harton has been bespoilt by the ugly scars of the coal mining industry, resulting in the slag-heap seen in the right background. In 1956 the South Shields to Sunderland service was still steam worked and on 14th August of the same year one of the local passenger trains is being hauled by a filthy G5 Class 0-4-4T No 67343 which is heading for its home base at Sunderland. (Peter Hay)

'9) As modernisation began to bite in the north-east, withdrawals of steam engines outstripped the capacity to scrap them and forlorn lines of locomotives began to gather at various depots. Amongst these was the shed at 55C Farnley Junction where snow lies on the ground on 5th January 1963. The focus of attention is on withdrawn LMS *Royal Scot* Class 4-6-0 No 46103 *Royal Scots Fusilier*, condemned from 55A Leeds (Holbeck) the precious month. (Alan Bailey)

80) The setting for this excellent photograph is West Hartlepool shed yard on 11th June 1949, some eighteen months after nationalisation. In the left background are well-worn steps leading to the ancient coaling stage and in the right of the frame is a small overhead gantry of lower quadrant signals. Centrepiece of this print is J94 Class 0-6-0ST No (6)8046 and a member of the footplate crew who looks rather self-conscious as he is photographed. (A.N.H.Glover)

81) Superpower for a local passenger train at Leeds (City) station on 23rd July 1955. Standing in the sunlight is LMS Hughes Class 6P5F 'Crab' 2-6-0 No 42721, from 26B Agecroft, which is piloting an unidentified LMS *Jubilee* Class 4-6-0 which has a good head of steam. In the right of the picture is D49 Class 4-4-0 No 62759 *The Craven*, of 50D Starbeck, which is in charge of an express. This latter engine survived in service until January 1961. (B.K.B.Green)

2) By coincidence this next picture portrays another D49 Class 4-4-0 popularly known as 'Hunt's' with their connections (names) of famous foxhunts in England, although many were also named after British shires and counties. Departing from Hull station in excellent external condition is No 62767 *The Grove*, from 53B Botanic Gardens, which has the logo of its new master in the summer of 1950. Withdrawal came in October 1958 for this loco. (D.K.Jones)

3) Possibly fresh from overhaul, Gresley inspired J39 Class 0-6-0 No 64910 stands lifeless in a section of the yard at its home shed at 53A Hull (Dairycoates) in June 1959. No 64910 took its leave of Dairycoates depot in September 1960, moving on to 55H Leeds (Neville Hill) followed by a move to 52B Heaton in January 1961. One final transfer in August 1962 took No 64910 to 52D Tweedmouth. After withdrawal in December 1962 it was cut up at Cowlairs. (N.E.Preedy)

84) Another locomotive looking fresh from shops is B16/3 Class 4-6-0 No 61467, based at the local shed at 51A, as it threads a path through Darlington with a fitted freight on 28th March 1959. No 61467 is a Thompson rebuild (1944) of an earlier Raven designed engine. It was reallocated to 50B Hull (Dairycoates) in December 1962 and survived in revenue earning service there until May 1964 after which it was later scrapped at Drapers, Hull. (D.K.Jones)

85) J72 Class 0-6-0T No 68720 (52A Gateshead) is on station pilot duty at Newcastle (Central) on 24th August 1957. This North Eastern Region engine (E1 Class) was built in April 1922 by Armstrong Whitworth as NER No 2313 and was renumbered to 68720 in February 1950. With their small driving wheels the J72's were mostly employed as freight shunters. At a later period in time some were repainted in lined-out livery for use at main line stations. (N.L.Browne)

6) A trio of diesel locomotives occupy the tracks in the left of this picture as resident B1 Class 4-6-0 No 61002 *Impala* basks in the sunshine at the end of the vast yard at 50A York on a summer's day in 1962. Allocated to 50A for many years, *Impala* took its leave of the depot at York in June 1965, moving to a new abode at 50B Hull (Dairycoates). After withdrawal in June 1967 it was stored at 50D Goole prior to scrapping. (D.K.Jones)

37) Following overhaul at the nearby workshops J27 Class 0-6-0 No 65859 waits at 51A Darlington to be steamed and returned to its home shed at 5IG Haverton Hill on 22nd June 1952. Keeping No 65859 company is J39 Class 0-6-0 No 64758. The depot at Haverton Hill closed completely on 14th June 1959 and all but three of its allocation was transferred to 51L Thornaby. No 65859 had already been drafted there four months earlier. It survived in service until September 1966. (A.N.H.Glover)

88) Under partially clear signals Riddles War Department Class 8F 2-8-0 No 90230, from 56B Ardsley and in sparkling condition, has steam to spare as it glides past the signalbox at Witham Sidings, near Wakefield on 1st October 1964. Records show us that from January 1957, No 90230 was allocated to 51B Newport, 50B Hull (Dairycoates), 50A York and 56A Wakefield before moving to Ardsley. Its final sanctuary was at 51C West Hartlepool. (R.S.Carpenter)

89) A4 Class 4-6-2 No 60007 *Sir Nigel Gresley* is noted in steam in the yard at 52A Gateshead on 17th August 1963. Although sporting a 34A Kings Cross shedplate, *Sir Nigel Gresley* was in fact allocated to 34E New England by this date. Standing next to No 60007 is V3 Class 2-6-2T No 67691, the last of the class numerically speaking. It had been at 52A for some two months following a transfer from 'up the road' at 52B Heaton. (N.E.Preedy)

90 The former North Eastern Railway depot at Sunderland, which carried the codes of 54A and 52G under BR, consisted of two sheds. In the left background we can see the four-road straight shed and in the right of the frame we can make out part of the covered roundhouse. In the foreground, sporting a badly scorched smokebox door, is a resident of the shed, J27 Class 0-6-0 No 65865 on 12th December 1964. The shed closed in September 1967. (N.E.Preedy)

91) A swirl of smoke and steam is driven sidewards by a strong wind as Gresley V2 Class 2-6-2 No 60810 battles against the elements as it heads northbound near Darlington with a down express fitted freight on a drab, but unknown day in 1963. Allocated to 50A York, No 60810 had been there ever since a transfer took it away from 52B Heaton after many years. Condemned from York shed in November 1965, No 60810 was scrapped at Station Steel, Wath in 1966. (D.K.Jones)

92) 50C Selby shed consisted of two large roundhouses situated between the station and Hambleton. It was to fall early victim to closure (October 1959) and for some years thereafter its yard was used for storage purposes (mostly coal wagons). On 3rd March 1954 one of its allocation, LMS Class 4 'Flying Pig' 2-6-0 No 43123 is seen in the shed yard in the company of TI Class 4-8-0T No 69912 which was withdrawn in October 1959. (B.K.B.Green)

93) LMS Stanier Class 8F 2-8-0 No 48271, from 18A Toton, is on alien territory as it rattles along light engine at Milner Royd Junction, near Sowerby Bridge on 10th September 1960. No 48271, a longstanding resident of Toton, moved on to pastures new at 17B Burton in April 1963. A final transfer in August 1966 saw it moving to 8E Northwich from whence it was withdrawn in August 1967. It was cut up at Buttigiegs, Newport in May 1968. (D.K.Jones)

94) Another 'stranger-in-the-camp', this time the location is York station on a bright and sunny 2nd August 1957. BR 'Crosti-Boilered' Class 9F 2-10-0 No 92024, from 15A Wellingborough, is way off the beaten track as it heads southbound light engine. No 92024 remained faithful to Wellingborough until January 1964 when it moved on to 15C Kettering. Later moves saw it in service from the depots at 12A Carlisle (Kingmoor) and 8H Birkenhead. (D.K.Jones)

95) We complete this trio of 'foreign' locomotives at work and rest on the North Eastern Region with a photograph of 12A Carlisle (Kingmoor) based LMS Class 5 4-6-0 No 45466 by the antiquated coaling stage at 55F Bradford (Manningham) on 31st May 1961. This depot, of Midland Railway origin (20E), had been transferred under the control of the North Eastern Region operating authorities early in 1957. It closed completely in May 1967. (D.K.Jones)

96) A trio of bedraggled looking LMS Ivatt Class 4 'Flying Pig' 2-6-0's are all out of steam outside the north end of the straight running shed at 51A Darlington on 4th September 1965. In addition to this shed there was also a small roundhouse used to stable examples of Darlington's small tank engine fleet. Nearest to the camera in this photograph is No 43050 which was to soldier on at 51A until closure to steam on 26th March 1966. (F.Hornby)

97) 1949 built (Doncaster Works) A1 Class 4-6-2 No 60149 *Amadis*, of 36A Doncaster, pounds towards the camera at Cardigan Road sidings, Leeds on 8th August 1959. *Amadis* is in charge of a Liverpool to Newcastle express, soon to be dominated by diesel traction. Once of 34A Kings Cross, No 60149 had been at 36A since October 1958. It remained there until taken out of traffic in June 1964, being scrapped at Wards, Killamarsh early in 1965. (M.Joyce)

98) The former North Eastern Railway Q6 Class 0-8-0's were used almost exclusively on freight duties in and around the north east of England. Designed by Raven they were first introduced into service in 1913 and some examples survived until September 1967. On 28th September 1963 No 63363, from 52H Tyne Dock, is seen near to the station of the same name with a mineral train. One example of the class, No 63395 is preserved. (N.E.Preedy)

99) V3 Class 2-6-2T No 67646 is seen out of steam at its home shed at 52B Heaton in May 1957. Allocated to Heaton shed for many years it moved 'down the road' to 52A Gateshead in June 1963. It was one of the last members of the class to be withdrawn, in November 1964, three months after the closure of Heaton shed. On the occasion of the author's last visit to Heaton, on 6th April 1963, there were twenty-one steam engines to be seen. (N.E.Preedy)

100) Although the former LMS shed at Royston (20C) became the property of the North Eastern Region early in 1957 (55D) no locomotive of LNER origin was allocated to the depot right up until closure in November 1967. Most of Royston's allocation consisted of freight orientated types like LMS Class 4F 0-6-0 No 43906 (withdrawn in November 1965) and LMS Class 8F 2-8-0 No 48710 (withdrawn in October 1967). (T.R.Amos)

101) Washington (County Durham) saw very few passenger trains, being situated on several busy freight routes. On 12th August 1956, V3 Class 2-6-2T No 67688 (52A Gateshead) arrives bunker-first with the empty stock which will form the tea-time workers train to Newcastle. This locomotive was the subject of a 'paper transfer' to 64B Haymarket in December 1962, the same month as withdrawal. It was cut up at Darlington Works in June 1963. (Peter Hay)

02) A small number of upper quadrant and 'dummy' signals are on parade at Durham station as A1 Class 4-6-2 No 60137 *Redgauntlet*, from 52B Heaton, roars through with an empty stock working on 11th August 1961. Built at Doncaster Works in 1948, *Redgauntlet*, a rarely photographed engine, was taken out of traffic from 52D Tweedmouth in October 1962. It was stored at 52C Blaydon for several months before being scrapped at Doncaster. (D.K.Jones)

103) West Hartlepool depot (51C) consisted of a twin roundhouse and a separate three-road straight shed. It was normally associated with being an outpost for a selection of former North Eastern Railway locomotives. It also had on its books a small number of LMS Class 4 'Flying Pig' 2-6-0's like No 43128, seen here in bright sunshine in the yard facing WD Class 8F 2-8-0 No 90067 on 4th September 1967. It was withdrawn in July 1965. (N.L.Browne)

104) As railway track workers sort out a problem in the left of this photograph, a member of the footplate crew of the compact, but powerful locomotive on view looks towards the camera and has his picture taken for posterity. The locomotive concerned is J27 Class 0-6-0 No 65827 seen in bright sunshine in the shed yard at 50A York on 25th May 1952. Seven years on and No 65827 was no longer with us, being withdrawn from 50F Malton. (N.E.Preedy)

105) The graceful lines of 1949 built A1 Class 4-6-2 No 60129 *Guy Mannering*, from 52B Heaton, are clear for all to see as it accelerates away from the sharply curved station at Hartlepool on an unspecified date in 1961 with an express duty. Once of 52A Gateshead, *Guy Mannering* had been at Heaton shed since August 1960. In September 1962 it moved on to a new abode at 52D Tweedmouth, followed by a further move to 50A York in July 1965. (N.E.Preedy)

06) LMS Stanier Class 8F 2-8-0 No 48348 is well away from its home base of 6B Mold Junction on the outskirts of Chester as it drifts by a row of stone-built terraced houses at Dewsbury with a mineral train on 5th August 1962. Transferred to 6C Birkenhead in February 1963, No 48348 remained there until June 1965 when it had a move to a final bolt-hole at 10F Rose Grove where it continued to work until the end of steam on BR in August 1968. (D.K.Jones)

07) More former LMS steam power, this time at 55C Farnley Junction, where three locomotives, condemned and no longer wanted, lie cold and still on a side road adjacent to the large running shed on 18th April 1963. From left to right are an unidentified LMS Class 3F 'Jinty' 0-6-0T, LMS Class 6P5F 'Crab' 2-6-0 No 42713, withdrawn from 55C in December 1962 and an unidentified LMS *Royal Scot* Class 4-6-0 complete with a sacked chimney. (D.K.Jones)

108) It is not often that one could see a WD Class 8F 2-8-0 at the head of a passenger train, let alone in sparkling condition, but this is the case for No 90348, from 56A Wakefield, which is in charge of the empty stock of a Railway Correspondence and Travel Society special at Darlington on 13th May 1962. All in all, No 90348 was allocated to Wakefield shed from June 1958 until October 1966 after which it was based at 52G Sunderland. (D.Harrison)

109) The wheels of industry are on the move at West Hartlepool on 11th August 1956 where the day is as drab and grey as begrimed J27 Class 0-6-0 No 65865, from the local shed at 51C, which is in charge of what appears to be a pick-up freight. An attempt has been made to smarten the area up with the planting of small bushes in between the tracks. No 65865 later worked from the sheds at 51L Thornaby, 52G Sunderland and 52F Blyth. (Peter Hay)

10) The clock on the face of the large water tower in the left of this picture informs us that it is almost fifteen minutes to four in the afternoon as D49 Class 4-4-0 No 62727 *The Quorn* (50D Starbeck) sizzles and steams in the yard at 50B Leeds (Neville Hill) on 3rd April 1955. Fitted with Lentz rotary cam valve gear and outside steampipes, *The Quorn*, withdrawn in January 1961, was named after a famous 'hunt' in Leicestershire. (F.Hornby)

11) The J72 Class 0-6-0 Tanks were designed by Worsdell and after being initially introduced in 1898 eventually came to a total of 113 units - BR Nos 68670-68754 and 69001-28. The class remained intact until 1958 and by January 1963 they had dwindled in number to just fourteen active members. Outside its home shed at 51A Darlington on 22nd June 1952 is No 68707 which served from 51C West Hartlepool later on in life until condemned in April 1962. (A.N.H.Glover)

112) The end for steam locomotion on the North Eastern Region was on the immediate horizon when this photograph was taken in August 1967 at 55A Leeds (Holbeck). In steam next to a stored sister engine is Holbeck based LMS *Jubilee* Class 4-6-0 No 45697 *Achilles* seen inside one of the large roundhouses. For many years a Carlisle (Kingmoor) steed, *Achilles* had found its way to Holbeck via 24E Blackpool and 27A Bank Hall. (R.Butterfield)

113) Two 'foreigners' are noted in the heart of North Eastern Region territory - circa 1964. LMS Class 5 4-6-0 No 45051, from far off 1H Northampton, pilots an unidentified LMS *Jubilee* Class 4-6-0 on a through track at York station with a southbound parcels train. Once of 3E Monument Lane, 3D Aston, 21B Bescot and 24J Lancaster (Green Ayre), No 45051 also served from 2A Rugby and 6D Shrewsbury prior to withdrawal in November 1966. (D.K.Jones)

14) The former London and North Western Railway shed at Farnley Junction, coded 25G by the London Midland Region from 1948-1956 and 55C by the North Eastern Region from 1957-1966, never possessed a turntable. Locomotives were turned via the use of a triangle within the close vicinity of the depot. BR Class 3 2-6-0 No 77001, an inhabitant of 55C, is probably taking advantage of this set-up as it steams backwards on 1st May 1964. (D.K.Jones)

5) Looking in ex.works condition, LMS Class 8F 2-8-0 No 48622, fitted with a small snowplough, is on show in the yard of its home shed at 55B Stourton in the company of a begrimed sister engine on 29th August 1965. In the background is the single covered roundhouse once owned by the Midland Railway. It came under the control of the North Eastern Region authorities at the end of 1956 and eventually closed its doors to steam in January 1967. (R.Picton)

116) Although it is some six and a half years since nationalisation the tender of the engine in the left of this photograph still carries the logo of the London North Eastern Railway somewhat of a rarity by this stage in time. The centrepiece of this study is of former North Eastern Railway J27 Class 0-6-0 No 65816 which is standing out of steam by a pile of discarded ash at 51C West Hartlepool, its home base on 13th June 1954. (B.W.L.Brooksbank)

117) A splendid panoramic view of the extensive trackwork and station at Stockton on a sunny day in 1961. Facing the camera is a light engine in the shape of Q6 Class 0-8-0 No 63402, from 52C Blaydon, destined for withdrawal in September 1964 from 52F Blyth. In the right background we can just make out Al Class 4-6-2 No 60140 *Balmoral*, of 50A York, which is departing with the 1.00pm express to Colchester, originating from York. (G.D.Appleyard)

18) Bright sunshine beats down on the quadruped tracks near to Mirfield, controlled by colour light signals, in the summer of 1965. Heading towards the camera is BR Class 4 4-6-0 No 75043, allocated to 8L Aintree, which is in charge of a five-coach 'express'. Once of 15D Bedford, 15C Leicester (Midland), 17A Derby and 8R Walton-on-the-Hill, No 75043 had arrived at Aintree shed in December 1963. Its final home was at 10A Carnforth. (R.Butterfield)

19) We make our departure from North Eastern Region metals with this study of former Great Central Railway O1 Class 2-8-0 No 63652 which is a visitor to 50A York from 41H Staveley GC on 6th June 1962. In January 1957, No 63652 was based at 38B Annesley and it was drafted to 31B March during the following month. Further moves took it to Staveley GC (twice) and 40E Colwick. Condemned from 41H in November 1963 it was scrapped in May 1964. (D.K.Jones)

120) As their charge waits impatiently the footplate crew of a rather less than clean BR Class 4 2-6-4T No 80124, from 62B Dundee Tay Bridge, awaits the guard's 'right-away' at Newport-on-Tay East station (closed in 1969) with the 6.00pm local passenger train bound for Tayport on 7th July 1964. A longstanding inmate of 62B, No 80124 was drafted to 64A St.Margarets (Edinburgh) in February 1966 where it survived until the end of the year. (A.F.Nisbet)

21) Fresh from overhaul at the not too distant Inverurie Works, former North British Railway D30 Class 3P 4-4-0 No 62421 *Laird o' Monkbarns* waits to be steamed again and returned to its its home shed at 64A St.Margarets (Edinburgh) in the yard at 61A Kittybrewster on 26th June 1955. *Laird o' Monkbarns* was destined to be one of the last active members of this class, known as 'Scotts', being withdrawn from St.Margarets in June 1960. (R.Butterfield)

22) For many years Fort William shed was home for a batch of ex. GNR K2 Class 2-6-0's equipped with side-window cabs and named after local Scottish lochs. One of their number, No 61774 *Loch Garry*, is noted on the turntable at 65J in the mid-fifties, sporting an intriguing 'bulge' at the base of the chimney - the purpose of which is unknown! Withdrawn from 65A Eastfield (Glasgow) in April 1958, *Loch Garry* was scrapped at Kilmarnock Works. (R.S.Carpenter)

123) The Great North of Scotland Railway was unique among Britain's private railway companies in using 4-4-0's for all main line passenger and freight duties and the twenty-one members of the D40 Class, with 6' 1" driving wheels, introduced in 1899, did yeoman service well into the fifties. On 25th June 1953 No 62272, heading a local freight at Elgin station, still looks in tip-top condition in the hot summer sunshine. (R.Butterfield)

124) Accompanied by a 350hp 0-6-0 diesel shunter, B1 Class 4-6-0 No 61347, from 61B Aberdeen (Ferryhill) and fitted with a small snowplough, is a rare visitor to the diesel depot at 61A Kittybrewster on a bright day in April 1964. Four months later No 61347 was despatched south to a new abode in Edinburgh at 64C Dalry Road. Withdrawn in April 1967, from 62A Thornton Junction, No 61347 was stored at 62B Dundee Tay Bridge prior to being scrapped. (N.E.Preedy)

25) Whilst on the subject of Edinburgh, we find ourselves in the shed yard at 64B Haymarket in the summer of 1957, where we espy an immaculate looking Gresley V2 Class 2-6-2 No 60932 in steam. No 60932 is a visitor to Haymarket from 52D Tweedmouth. In June 1958 it was drafted to 52B Heaton, where it remained until December 1962. A final transfer took No 60932 to 50A York, from whence it was taken out of service in May 1964. (N.E.Preedy)

26) On a grey day on 22nd April 1957 D30 Class 4-4-0 No 62418 *The Pirate*, from the local shed at 62A, shows a clean pair of heels as it makes a spirited departure from Thornton Junction with the 12.08pm local passenger to Crail which lost its local services in 1965. By this date in time there were twenty-three of these elderly locomotives still in active service. *The Pirate* was withdrawn from Thornton Junction in August 1959. (Peter Hay)

127) The station nameboard in the left of this picture proudly proclaims 'Inverkeithing for Rosyth', but in recent times the once proud naval dockyard has suffered badly in the defence cuts. Looking on the bright side the locomotive seen at Inverkeithing in 1965 is still with us today. A melee of menfolk gather on the platform to admire the presence of A4 Class 4-6-2 No 60019 *Bittern*, from 61B Aberdeen (Ferryhill), with an express. (D.K.Jones)

128) K4 Class 2-6-0 No 61996 *Lord of the Isles*, of 65A Eastfield (Glasgow), steams majestically through Princes Street Gardens prior to arriving at Edinburgh (Waverley) station with a local passenger train from Glasgow (Queen Street) in September 1958. In May of the following year, *Lord of the Isles* was transferred to 62A Thornton Junction where it continued to work until withdrawal in October 1961 after which it was cut up at Inverurie. (D.K.Jones)

29) A remote outpost for Scottish steam was the small market town at Hawick, situated on the Waverley route between Carlisle and Edinburgh, long since defunct. It also possessed a diminutive locomotive shed once the property of the North British Railway and coded 64G by British Railways. On an unknown day in February 1961 one of its former inhabitants, C16 Class 4-4-2T No 67489 (condemned this same month) is noted in store. (D.K.Jones)

30) 62B Dundee Tay Bridge shed provided a haven for a small number of A2 Class 4-6-2's during the latter years of steam and was a popular venue for spotters and photographers alike during the twilight years of the depot. On 4th March 1964 No 60528 *Tudor Minstrel*, built in 1948, is in steam in front of the running shed at 62B. Although not looking in the best of external condition *Tudor Minstrel* survived in active service until June 1966. (D.K.Jones)

131) It is the early hours of a drab spring morning and passengers mill around on the platform as a guard makes notes next to the fireman (complete with knotted handerchief) of A4 Class 4-6-2 No 60027 *Merlin* (65B St. Rollox) which is in charge of the 05.27 departure from Aberdeen to Glasgow (Buchanan Street) express on 30th May 1964. Once part of the proud fleet of A4 Pacifics based at 64B Haymarket, *Merlin* had moved to 65B in May 1962. (D.K.Jones)

132) Bearing the logo of its new master on the side tanks C15 Class 4-4-2T No 67454 looks in reasonable external condition despite being in store with a sacked chimney at 65C Parkhead on 3rd June 1951. These compact but strong locomotives were designed by W.P.Reid and between 1911 and 1913 a total of thirty were built for the North British Railway. By the end of 1956 only two members of the class were still in service, Nos 67460/74. (W.Boyden)

33) In pre-grouping days Keith was an important meeting place of the Highland and Great North of Scotland Railways and even as late as the 1950's the shed at 61C was still home to a mixed bag of nineteen steam locomotives. GNSR D40 Class 4-4-0 No 62268, photographed whilst having attention to its smokebox on 29th June 1955, was just one of a dozen of the class to be stationed there. It was withdrawn within the next eighteen months. (R.Butterfield)

34) We can clearly see the spartan surroundings at St.Combs in the north of Scotland, terminus of the branch line from Fraserburgh, on a wet and miserable day in the fifties. Running around its three-coach local at St Combs is LMS Class 2 2-6-0 No 46460, from 61A Kittybrewster and equipped with a cow-catcher. No 46460 was to remain in Scotland for the rest of its career prior to being withdrawn from 67C Ayr in August 1966. (R.Butterfield)

135) The Gresley V2 Class 2-6-2's were frequent performers on passenger and freight turns between Edinburgh and Aberdeen via the Forth and Tay bridges. Here we see 64A St.Margarets (Edinburgh) based No 60931 heading a Saturdays Only Aberdeen to Edinburgh (Waverley) relief express through Dundee's long-closed (1939) and bedgraggled looking Esplanade station on 24th August 1963. It was condemned from 64A in September 1965. (A.F.Nisbet)

136) The diminutive former North British Railway Y9 Class 0-4-0 Saddle Tanks were first introduced into service in 1882 from a design of Holmes. Many were equipped with small wooden carts as 'tenders' and wooden bufferbeams. Most were to be found on sharply curved lines in dockland areas and were often hired out to private contractors. On 13th August 1948, No 6(8122) is found in residence at 64A St. Margarets (Edinburgh). (B.W.L.Brooksbank)

37) Up until May 1959 all five members of the K4 Class 2-6-0's were allocated to 65A Eastfield (Glasgow) primarily for use on the West Highland Line. As more and more diesels came onto the scene they were gradually transferred away to a new haven at 62A Thornton Junction and all were in residence by the end of the year. On 12th September 1959 No 61993 *Loch Long* and its footplate crew were photographed at 62A. Note the etchings on the bufferbeams. (W.Boyden)

38) Sunlight and shadows at Princes Street Gardens in the magnificent city of Edinburgh in the summer of 1960. The focus of the photographer's attention is the final member of the A1 Class 4-6-2's, No 60162 *Saint Johnstoun*, from 64B Haymarket. A numerical batch of these engines, Nos 60159-62 were all based at 64B until September 1963 when Haymarket closed to steam. They moved to 64A St.Margarets and all were condemned by the end of 1963. (N.E.Preedy)

139) Another famous class of 4-4-0's based at sheds in Scotland were the former North British Railway D34 'Glen's'. Introduced in 1913 some lingered on until December 1961. On a warm day in August 1953, No 62470 *Glen Roy* is seen in the yard of its home shed at 65A Eastfield (Glasgow) in the company of 64A St.Margarets based K3 Class 2-6-0 No 61857. After being withdrawn in May 1959, *Glen Roy* was scrapped at Connells, Calder in February 1960. (N.E.Preedy)

140) As withdrawals of redundant steam locomotives in Scotland began to exceed the capacity to cut them up, several 'dumps' were created. The most infamous was at 64F Bathgate, between Edinburgh and Glasgow. Although looking for all the world that the end is nigh, V2 Class 2-6-2 No 60931, seen at Bathgate on 10th June 1962, was later reinstated and returned to 64A St.Margarets. Also in the frame is V3 Class 2-6-2T No 67615 (64B Haymarket). (F.Hornby)

41) On an unknown date in 1955, 1925 built (Doncaster Works) A3 Class 4-6-2 No 60057 *Ormonde* passes its home shed at 64B Haymarket with a northbound up express from Edinburgh (Waverley) to Aberdeen consisting of a mixed rake of coaching stock. Modified with a double chimney (July 1958) and German style smoke deflectors in September 1961, *Ormonde* was to survive in active service until October 1963 being withdrawn from 64A St.Margarets. (N.E.Preedy)

42) The J37 Class 0-6-0's, of North British Railway origin, were employed on freight duties for the majority of their working lives and were a common sight over many routes in Scotland. On an overcast day in 1965 a 'bulled-up' member of the class, No 64547, from 62B Dundee Tay Bridge, is seen light engine at Tay Bridge station. Allocated to 64A St. Margarets for many years, No 64547 made the move north to Tay Bridge shed in March 1964. (R.S.Carpenter)

143) An impressive church occupies the background of this picture taken at Fraserburgh in the early fifties. Adjacent to the small station at Fraserburgh is the compact two-road shed, a sub-depot of 61A Kittybrewster. Standing in the bright sunlight is Bl Class 4-6-0 No 61324, of 61A. After the closure of Kittybrewster to steam in the summer of 1961, No 61324 was drafted to 12C Carlisle (Canal). Withdrawal came in November 1965. (R.Butterfield)

144) Seconded from 62A Thornton Junction in the Kingdom of Fife, K4 Class 2-6-0 No 61995 *Cameron of Lochiel* is back on its old stamping ground on the West Highland line in July 1960. *Cameron of Lochiel* is seen at Rannoch station during a photostop whilst working a Stephenson Locomotive Society special. Withdrawn from Thornton Junction in October 1961, No 61995 was scrapped at the Halbeath Wagon Works, Dunfermline in March 1962. (N.E.Preedy)

45) Quite obviously there is time to spare before ex. Great North of Scotland Railway D40 Class 4-4-0 No 62267 (61C Keith) departs from Elgin station with its train of Gresley LNER coaching stock on 25th June 1953. Note the outward-pointing oil lamp mounted by the cab windows of No 62267 and the ancient and bedraggled six-wheeled van in the right background. The last survivor of the class, No 62277 *Gordon Highlander*, is preserved. (R.Butterfield)

46) Named after one of the characters in the 'Scott' novels, D29 Class 4-4-0 No 62401 *Dandie Dinmont*, minus shedplate, stands amidst the weeds in the shed yard at 62A Thornton Junction on 5th June 1949, surrounded by WD Class 8F locomotives. Introduced into service in 1909, from a design by W.P.Reid, some examples had already been withdrawn by the date of this picture. The survivors of the class were rendered extinct before 1955. (A.N.H.Glover)

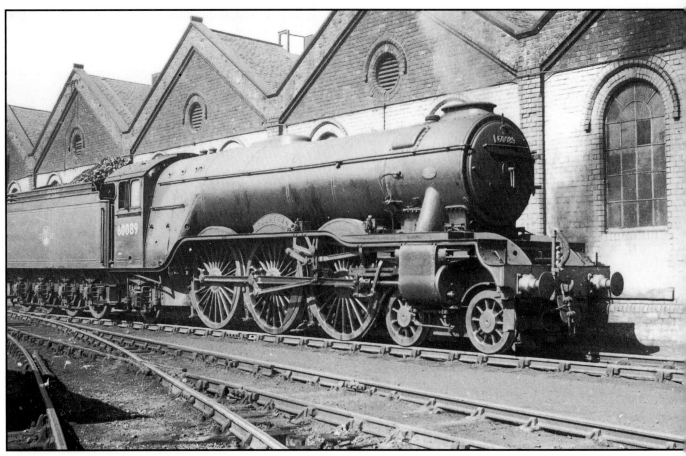

147) Looking every inch a pedigree, 1928 built A3 Class 4-6-2 No 60089 *Felstead*, equipped with a double chimney in October 1959, stands in steam alongside the running shed at its home base of 64B Haymarket on 21st May 1960. Rendered redundant by diesel power at 64B in December 1960, *Felstead* made the short journey across Edinburgh to a new home at 64A St. Margarets. Fitted with German smoke deflectors in October 1961 it was withdrawn in October 1963. (N.E.Preedy)

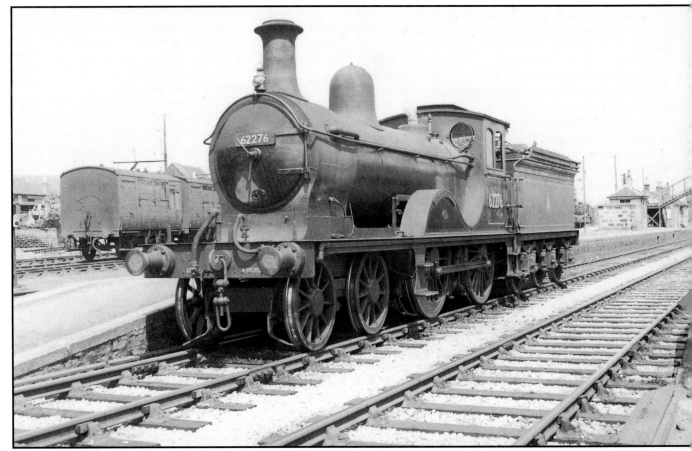

148) Maud station, closed in 1965, was the junction for the lines to the two fishing ports of Peterhead and Fraserburgh and on 23rd June 1953 still boasted a respectable goods yard. Former Great North of Scotland Railway D40 Class 4-4-0 No 62276 *Andrew Bain*, from 61A Kittybrewster and sporting a partially scorched smokebox door, is photographed at Maud Junction after arrival from Peterhead with a local passenger working. (R.Butterfield)

9) Having descended from his lofty position in the comfort of his signalbox, the signalman at Tayport prepares to take the single line token from the footplate crew of BR Class 4 2-6-4T No 80123 under the watchful eyes of a group of youngsters as it arrives bunker-first at the head of the 6.27pm local passenger from Dundee Tay Bridge. Allocated to Tay Bridge shed, No 80123 moved on to pastures new at 66A Polmadie (Glasgow) in March 1965. (A.F.Nisbet)

50) The K2 Class 2-6-0's were a modification of ten earlier locomotives of Class K1, both of which were designed by Sir Nigel Gresley. The K2's were distinguishable from the K1's by virtue of having outside steampipes. Their most famed work was over the West Highland route and the locos associated with the same were given names and side-window cabs. In October 1952, No 61787 *Loch Quoich*, from 65A Eastfield, is seen on Mallaig shed. (R.Butterfield)

151) A3 Class 4-6-2 No 60069 *Sceptre*, from 52B Heaton, gracefully glides past Portobello signalbox in Edinburgh with an unidentified Anglo-Scottish express in September 1957. Modified in August 1959 with a double chimney, *Sceptre* was one of only a handful of the A3's not to carry German style smoke deflectors. Between 1957 and withdrawal in October 1962, No 60069 remained loyal to the sheds in the North Eastern Region of BR. (N.E.Preedy)

152) BR Class 2 2-6-0 No 78046 is seen on a local freight duty at its home base of Hawick in June 1964. This was one of a series of sixty-five engines which first appeared in 1953 under the direction of R.A.Riddles. Like all of the BR classes this one was to be rendered extinct well before its economic life had expired, thanks to the 'rush to modernise' policy. No 78046 was condemned from 64A St.Margarets in November 1966. (N.E.Preedy)

53) Pacific power at rest on the turntable at 64B Haymarket on an overcast day in July 1958. Showing its fine lines is locally based A2 Class 4-6-2 No 60534 *Irish Elegance* (built at Doncaster in 1948). Ousted by the ever encroaching legions of mainline diesels, *Irish Elegance* was transferred to 64A St.Margarets in November 1961. Just over twelve months later it was condemned from 64A and cut up eventually at Campbells, Airdrie. (N.E.Preedy)

54) On the subject of the shed at 64A St.Margarets we find ourselves in the yard of the same, where a begrimed A3 Class 4-6-2 merges in with the grim surroundings in September 1960. No 60041 *Salmon Trout*, its identifying features almost obliterated, had moved to 64A from 64B Haymarket two months earlier. Fitted with a double chimney (July 1959) it was equipped with the German smoke deflectors in January 1963. Withdrawal came in November 1965. (N.E.Preedy)

155) One of the lower quadrant signals in the left of this photograph is dangling at 'half-mast', as though the signalman does not have sufficient strength to set it in the correct 'off' position. As a member of the footplate crew looks towards the camera, Vl Class 2-6-2T No 67602, from 65A Eastfield (Glasgow), glides along light engine in readiness to take up a local passenger working at Balloch on the shores of Loch Lomond on 21st June 1955. (R.Butterfield)

156) A large gasometer looks down upon the shed scene at 65E Kipps on a drab and grey day on 15th June 1958 where resident former North British Railway J83 Class 0-6-0T No 68461 (condemned) is seen in the company of its replacement, 350hp 0-6-0 diesel shunter No D3409. Closed at the end of 1962, most of the surviving steam engines were sent for scrapping. The depot was used for many years after closure for storage purposes. (F.Hornby)

57) Many years before the depot was used as a storage point for a vast number of redundant engines, 64F Bathgate housed one of the elegant, but elderly, D30 Class 4-4-0's. On an unknown date in the mid-fifties, No 62439 *Father Ambrose* stands lifeless in the shed yard near to the water tower. Condemned from Bathgate in September 1959, *Father Ambrose* was later despatched to the Motherwell Machinery & Scrap Company for cutting up. (D.K.Jones)

58) Especially cleaned up for the occasion J36 Class 0-6-0 No 65345, of 62A Thornton Junction, has been purloined to work a Scottish Railway Preservation Society Railtour, seen at Mussleborough on 27th August 1966. Despite coming from a class which first appeared on the railway scene in 1888, Nos 65288 (62C Dunfermline) and 65345 had the dubious distinction of being the last two working steam locos in Scotland, being withdrawn in June 1967. (N.E.Preedy)

159) We can just catch a glimpse of the waters of the Firth of Tay in the distance as the driver of work-stained BR Class 4 2-6-4T No 80123, from 62B Dundee Tay Bridge, has a chat with the Tay Bridge South signalman under the watchful eye of his fireman prior to setting off from Wormit station for Dundee Tay Bridge station with a late afternoon local passenger train. Wormit, of North British origin, closed to passengers in 1969. (A.F.Nisbet)

160) Unlike most motive power depots in Britain, which were owned by their respective companies, Ferryhill (Aberdeen) was jointly owned by the Caledonian and North British Railways. It had a mixed allocation of locomotives, from the mighty Pacifics down to the humblest of tank engine types. One of its inhabitants, V2 Class 2-6-2 No 60898 is ready for the road in the yard at 61B in March 1961 some two years plus away from withdrawal. (N.E.Preedy)

61) Although fully coaled a brace of former North British Railway D34 'Glen' Class 4-4-0's are stored tender to tender on a side road at 64E Polmont on 17th May 1959. Nearest the camera is No 62471 *Glen Falloch* behind which is No 62488 *Glen Aladale*. Both locomotives, allocated to 64A St.Margarets (Edinburgh), were later restored to traffic, but not for long. Both were withdrawn during 1960 and cut up later in the same year. (J.M.Tolson)

62) Once part of the pride of the fleet of steam locomotives allocated to the 'Top-Link' shed at 34A Kings Cross, A4 Class 4-6-2 No 60006 *Sir Ralph Wedgwood* had made the move north of the border to 64A St. Margarets in October 1963. On 9th June 1965, whilst based at 61B Ferryhill (Aberdeen), *Sir Ralph Wedgwood* has been relegated to hauling a Class 8 loose-coupled freight, seen here passing Monifieth station between Arbroath and Dundee. (K.L.Seal)

163) Scottish based steam locomotives were often more distinguishable from their English and Welsh counterparts by virtue of the fact that the cabside numbers were of a larger format. This can clearly be seen on the cabside of BR Class 5 4-6-0 No 73120, from 63A Perth, noted in excellent external condition at Edinburgh (Waverley) station on 16th June 1957. No 73120 made its departure from Perth shed in December 1962, moving to 67A Corkerhill. (N.E.Preedy)

164) From 1962 to 1964 most of the Glasgow to Aberdeen expresses were in the hands of ex. LNER Pacifics, but on 30th March 1964 the 5.30pm from Aberdeen to Glasgow (Buchanan Street) service is in the quite capable, but less powerful, LMS Class 5 4-6-0 No 44703, fitted with a small snowplough. At this stage in time there were just three of these engines allocated to 61B Ferryhill, No 44703 being one of them, along with Nos 44794 and 45162. (D.K.Jones)

165) The only things living in this picture, apart from the photographer, are the weeds in the foreground and the trees in the background. All is cold and still on the railway front in the scrapyard at Inverurie Works on 24th June 1953. Hemmed in by two partially stripped locomotives is D40 Class 4-4-0 No 62242. This was an inside cylinder locomotive designed by W.Pickersgill and the class was constructed between 1899 and 1920. (R.Butterfield)

166) Looking in dire need of an overhaul, locally based V3 Class 2-6-2T No 67611 is in steam in the shed yard at 65C Parkhead as it awaits its next local passenger turn on 27th August 1957. Both variations of this class, V1 & V3, were identical except for the fact that the V3's had a slightly higher boiler pressure of 200psi. No 67611, withdrawn from active service at Parkhead in December 1962 was stored for many months at Bo'Ness prior to scrapping. (N.L.Browne)

167) After a photographic session a gaggle of enthusiasts make their way back to their seats in the carriages of this Railway Correspondence and Travel Society special situated at Middle West, Edinburgh on 20th September 1964. Heading the special with steam to spare is the now long preserved A4 Class 4-6-2 No 60009 *Union of South Africa*, from 61B Ferryhill (Aberdeen). Once of 64B Haymarket, No 60009 was taken out of normal service in June 1966. (D.Webster)

168) Freight engine power on show immediately outside the massive running shed at 65A Eastfield (Glasgow) on 25th August 1957 where one of its massed ranks of resident J37 Class 0-6-0's, No 64580, is seen in light steam with a fully stocked tender. In steam days Eastfield had a huge steam population supplemented by locomotives awaiting repair or fresh from the nearby works at Cowlairs after overhaul. 65A closed to steam in November 1966. (N.L.Browne)

69) Steam rises from the safety valves of V2 Class 2-6-2 No 60836, fitted with outside steampipes, as it stands near to the coaling stage at its home depot of 62B Dundee Tay Bridge on a sunny and warm looking 28th August 1965. No 60836 had been at 62B since a move from 61B Ferryhill in April 1964. Condemned in December 1966, No 60836 was stored at 64A St. Margarets and 62A Thornton Junction until August 1967 before being cut up. (N.E.Preedy)

70) The parent shed of 62C Dunfermline had two sub-sheds under its control, at Alloa and Kelty. The former was a small two-road open-ended affair which is clear for all to see in this photograph taken in 1950. Peeping out of the one end of the shed are two J38 Class 0-6-0's, Nos 65918 and 65934, both from 64A St. Margarets (Edinburgh). Both of these locomotives were withdrawn from 62C Dunfermline in late 1966. (R.Butterfield)

171) Possibly deputising for a Type 4 diesel or a Pacific, V2 Class 2-6-2 No 60951 passes through Edinburgh's Haymarket station with an Aberdeen express in June 1960. No 60951, a 64A St.Margarets locomotive, had been at the depot since April of this same year. This engine spent much of its working life based at both Haymarket and St. Margarets sheds before early withdrawal in December 1962. It was cut up at Cowlairs Works in October 1963. (N.E.Preedy)

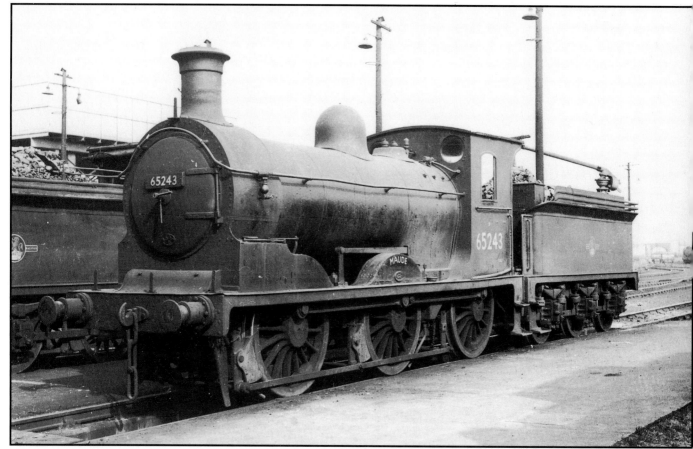

172) Former North British Railway J36 Class 0-6-0 No 65243 *Maude* is a visitor to 64F Bathgate from 64B Haymarket and is noted in the shed yard at 64F on 21st August 1959. Despite the diesel invasion of the late fifties/early sixties, *Maude* remained on the books at 64B until the depot closed to steam in the autumn of 1963. After stints at 64C Dairy Road and 64F Bathgate it was withdrawn from the latter for preservation in July 1966. (N.E.Preedy)

73) By coincidence this next portrait is of a now preserved former Scottish Region based locomotive. Once the pride of 61B Ferryhill (Aberdeen), A2 Class 4-6-2 No 60532 *Blue Peter*, fitted with a double blastpipe and chimney, poses for the camera on the turntable at 61B on an unknown date in 1960. Drafted to 62B Dundee Tay Bridge in June 1961, *Blue Peter* remained there until December 1966 when it returned to 61B, being withdrawn the same month. (D.K.Jones)

74) A pall of grey and black smoke erupts from the funnel of BR Class 4 2-6-4T No 80090, from 62B Dundee Tay Bridge, as it makes a spirited departure on single track from Tayport station on 6th July 1964. No 80090 is in charge of the 9.11am local passenger train to Tay Bridge. Once of 6H Bangor and 6A Chester, No 80090 was drafted to Scotland (62B) in March 1960. Withdrawn in March 1965 it was cut up at a yard in Faslane two months later. (A.F.Nisbet)

175) Former Caledonian Railway Class 2P No 55185, from 61C Keith, fusses around in the yard at 61A Kittybrewster in Aberdeen on 23rd June 1953. Note the partially scorched smokebox. 61A had a small number of LMS engines on its books, along with several BR types to boost the allocation of ex. LNER engines, but as far as the author knows it never owned any Caledonian locomotives. No 55185 was taken out of traffic from Keith in July 1961. (R.Butterfield)

176) For some obscure reason the mighty War Department Class 8F 2-10-0's based in Scotland were rarely photographed. One of their number, No 90760, a visitor to 62C Dunfermline from 66B Motherwell, lies dead in the shed yard on 16th June 1958. Of the twenty-five members of the class all but two examples were subjected to the mass withdrawals of 1962. No 90760 was condemned in May of this year and scrapped at Cowlairs Works. (N.L.Browne)

77) As is well known the depot at 64A St. Margarets (Edinburgh) was split into two sections either side of the East Coast Main Line. The smaller of the two had an open 'roundhouse' used to accomodate tank engine types like former NBR J88 Class 0-6-0T No 68348 seen here on 25th August 1957. It was withdrawn exactly twelve months later and scrapped at Kilmarnock Works. Keeping company with No 68348 is ex. NBR J83 Class 0-6-0T No 68450. (N.L.Browne)

78) In steam days millions of tons of freight were carried the length and breadth of Britain, mostly in loose-coupled wagons. Sadly, this is mostly a thing of the past as is begrimed former North British Railway J35 Class 0-6-0 No 64488, based at the near-at-hand shed (62A) as it negotiates a cross-over with a mineral train at Thornton Junction station on a drab day in 1958. It was condemned from 62A in October 1961 and scrapped two months late (Eric Light)

179) A rarity on British Railways was the building of twenty-eight members of the J72 Class 0-6-0 Tanks by BR to an unaltered former LNER design. They were numbered 69001-28 and constructed during the period from 1949-51. Most were allocated to depots on the North Eastern Region, but a few found their way to Scotland, like No 69015 seen on shed at 65C Parkhead on 23rd June 1956. It was withdrawn in September 1961 but not scrapped until 1963 (A.N.H.Glover)

180) We take our leave of BR STEAMING ON THE EX. LNER LINES - Vol. 4 with this photograph of a rather less than clean Thompson B1 Class 4-6-0 No 61102, from 62B Dundee Tay Bridge, as it departs from Tayport station in August 1962. Introduced into service in 1942, they eventually came to a total of 410 engines. The first to be withdrawn was No 61057 which was damaged beyond repair in an accident in 1950. Many engines were named after species of antelopes. (A.F.Nisbet)